My

I bite my food
with my teeth,

and chew it into little bits.

It goes into
my gut.

Here it is made smaller and smaller so . . .

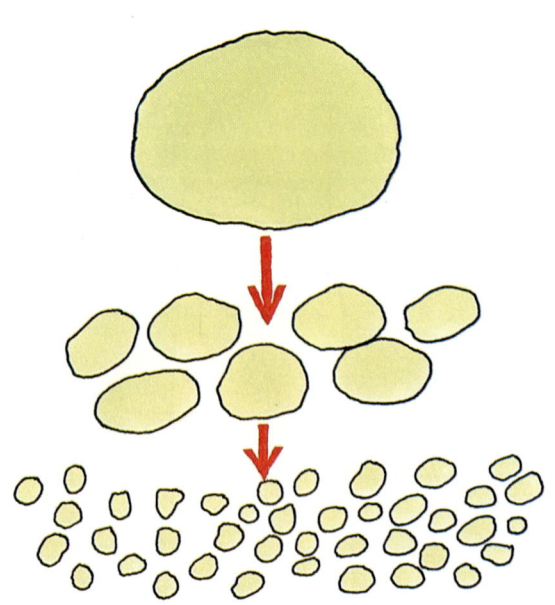

it can get into
my blood.

My blood
takes the
food to
all my body.

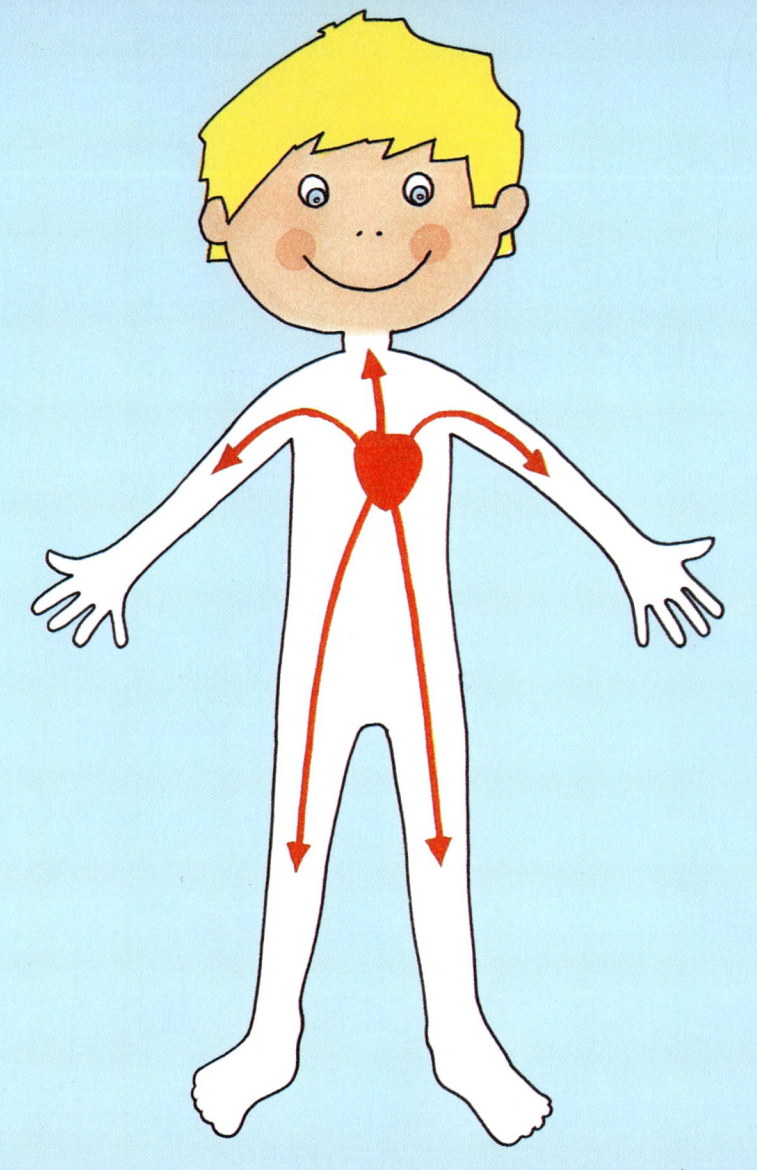

Food makes me grow big,

and helps me to run and jump.

Too much food can make me fat.

Too little food will make me thin.

The food I cannot use comes out when I sit on the toilet.